Early in the nineteenth century, as they turned to the chemical analysis of themselves and other living creatures, chemists found that most of the solid matter from which we are made has a remarkably uniform composition. Only a few elements are present and they occur in the proportions 54% carbon, 7% hydrogen, 16% nitrogen, 22% oxygen and 1% sulphur. Substances with this composition make up some 80% of the dry weight of muscle, 70% of the dry weight of skin and 90% of the dry weight of blood. Clearly they are of great importance in the construction and chemical activity of living organisms and the pioneer biochemists called them proteins, from the Greek for 'holding first place'.

After many years of study and a great deal of argument proteins are now known to be very large molecules with quite definite chemical structures. These molecules vary remarkably in their shapes and properties. The molecules of hair, skin, silk, tendons and bone, for example, are long and thin though they are still small, of course, by comparison with the size of things that we can see with the naked eye. Typically a molecule of collagen (the protein of tendon and bone) is roughly cylindrical, 280 nm long and only 1·4 nm in diameter: some 300 000 of these molecules joined end to end make up the length of a 100 mm tendon and roughly 10^{12} rows like this are needed to make up a tendon 5 mm in diameter. Such are the *fibrous* proteins, the main structural materials from which we are made. In contrast the protein molecules that are involved in most of the chemical activity within our bodies are more compact in shape, though they are found in a wide range of sizes. These *globular* proteins include haemoglobin, the red substance responsible for carrying oxygen in the blood; antibodies, the molecules that combine with foreign molecules and help to remove them from our bodies; some of the chemical messengers or hormones, such as insulin; and most important, the enzymes, the organic catalysts that control the rates at which the various chemical reactions in our bodies take place.

Amino acids and polymerization

These versatile proteins are chemically rather simple. They are polymers in which small molecules, the L-α-amino acids, of twenty different kinds are joined together in long chains. The amino acids all have the same general structure

$$H$$
$$|$$
$$R—C_\alpha—COOH$$
$$|$$
$$NH_2$$

in which a central carbon atom (called by convention the α-carbon atom, C_α) is connected to each of four different atoms or groups of atoms: a hydrogen atom (—H); a carboxylic acid group (—COOH); an amino group (—NH$_2$); and one of twenty different side chains of various chemical types (—R). Under normal physiological conditions the amino acids exist in a doubly ionized (dipolar) form. The carboxyl group ionizes forming COO⁻ and H⁺; the amino group, in which the nitrogen atom has two covalent bonds with the hydrogen atoms and a third with C_α, forms a dative bond between the remaining two electrons of the nitrogen valency shell and a proton, forming NH_3^+.

$$H$$
$$|$$
$$R—C_\alpha—COO^-$$
$$|$$
$$NH_3^+$$

Having atomic groupings of both kinds, they exhibit properties of both carboxylic acids and amines.

When four atoms are bonded to a carbon atom in this way they are arranged in space at the vertices of a tetrahedron with the carbon atom at the centre. Two different arrangements or *configurations* are possible, one of which is the mirror image of the other, and only one of these configurations is found in the amino acids that are constituents of protein molecules. They are all of the type known as L-amino acids which have the three-dimensional structure:

Viewed along the bond from H to C_α the other groups appear in the clockwise order CO.R.N, which is easy enough to remember.

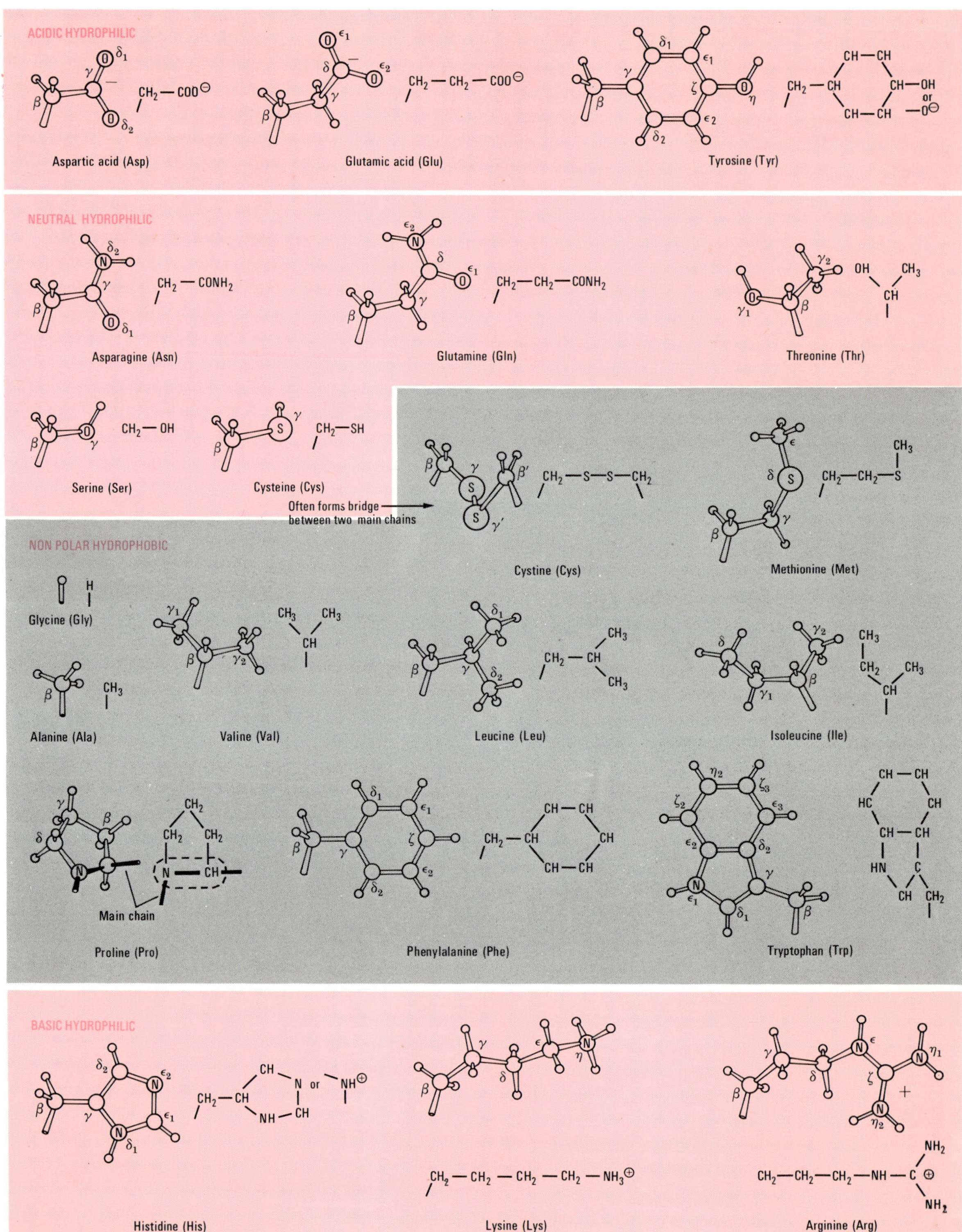

FIG. 1. The 20 naturally occurring L-α-amino acids.

4

The twenty different amino acids are distinguished by the R-groups or *side chains* shown in Fig. 1 where they are grouped according to their chemical types. The simplest of all is glycine in which R is another hydrogen atom (so that only one configuration of this molecule is possible). The other side chains include varying numbers of carbon, nitrogen, oxygen, sulphur (and hydrogen) atoms, which are often identified by the letters of the Greek alphabet, starting with $C\beta$. Thus the side chain of alanine is simply a methyl group ($-CH_3$) in the β-position while valine, isoleucine, and leucine have alkyl side chains, composed of three or four carbon atoms (with associated hydrogens) branched at the β- or γ-carbon atoms. In addition to these with non-polar side chains, there are amino acids with polar, acidic, basic, aromatic and sulphur-containing side chains and this makes possible a wide range of properties in the protein molecules that are made from them.

Linking two of these amino acid molecules together in the way that is characteristic of proteins involves only a simple condensation reaction in which the elements of water are eliminated and a peptide bond is formed.

$$\underset{H}{\overset{R}{H_3^+N-C_\alpha-COO^-}} + \underset{H}{\overset{R'}{H_3^+N-C_\alpha-COO^-}}$$

$$\downarrow$$

$$\underset{H\quad O}{\overset{R\quad\quad H\ R'}{H_3^+N-C_\alpha-\underset{\underbrace{\quad\quad\quad}_{\text{peptide bond}}}{C'}-N\quad C_\alpha-COO^-}} + H_2O$$

The resultant molecule is called a dipeptide and the parts of it that are derived from the original amino acids, with their characteristic side chains R and R' are known now as *amino-acid residues*. At one end an amino group survives and at the other a free carboxyl. Tripeptides consist of three amino-acid residues linked together by two peptide bonds, tetra-peptides and penta-peptides comprise four and five residues respectively and polypeptides are chains of very many amino-acid residues with each one linked to the next by a peptide bond. Individual protein molecules are composed of one or more polypeptide chains in each of which there may be several hundred amino-acid residues in a particular order known as the *amino-acid sequence*.

Now we see how the extraordinary versatility of proteins arises. Given building blocks of twenty different kinds that can be chosen in any relative numbers and arranged in any order in a linear sequence there is clearly an inconceivably large number of possible different molecules. Suppose that we are restricted to polypeptide chains of only 100 amino-acid residues, which is almost too small a number for a true protein. Each residue can be chosen in twenty ways so that 20^{100} different kinds of protein molecule of this size could be made, a number that is far greater than the supposed number of atoms in the whole universe.

Amino-acid sequence in insulin
For many years there was doubt and confusion about the nature of proteins, which were thought to be different in some vital way from other chemicals, but these arguments were resolved finally by Sanger who showed that insulin (a rather small peptide hormone) has a definite chemical structure with the amino-acid residues arranged in a fixed sequence. The molecules of insulin produced in the pancreas of a pig have the structure shown in Fig. 2 where the amino-acid residues are denoted

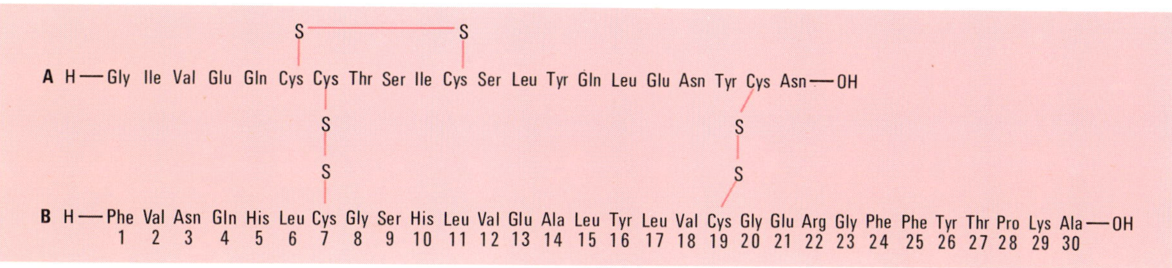

FIG. 2. The amino-acid sequence of pig insulin. Each molecule contains two polypeptide chains (A and B), disulphide bridges join the A and B chains together, and a third one connects different parts of the A chain.

by their standard three-letter abbreviations. This molecule is made from two short polypeptide chains and they are linked together by disulphide bridges, (—S—S—) that are formed with the elimination of hydrogen between the side chains of two cysteine residues (in cysteine, $R = —CH_2—SH$, see Fig. 1). Different proteins have different but equally definite sequences and many of these have now been determined by the methods pioneered by Sanger, Moore and Stein and others.

Three-dimensional structures of proteins

But the properties of a protein molecule depend not only on what groups of atoms are present in it but also upon how they are arranged in space. The three-dimensional structures of these molecules have therefore been studied intensively and we now know a number of them in detail. A most important principle has also emerged, that the shape or *conformation* adopted by a protein molecule depends only upon its amino-acid sequence. If this is the case, it should be possible to predict the folded conformation of a protein molecule from its chemical formula, but this has not yet been achieved despite a great deal of work. Nevertheless, some progress has been made and this can be said to have begun with the X-ray crystallographic studies, mainly by Pauling and Corey and their colleagues, of the three-dimensional structures of individual amino acids and simple peptides. A result typical of these analyses is given in Fig. 3 which shows the molecular structure of glycyl-L-asparagine ($^+NH_3 . CH_2 . CO . NH . CH(CH_2 . CO . NH_2)COO^-$) and the interactions between these molecules in the crystalline state.

From a study of the molecular dimensions and interactions revealed by these crystallographic studies, Pauling and Corey were able to suggest a number of conditions that are usually satisfied in the folding of a polypeptide chain. These are: (a) that the peptide groups linking two amino-acid residues

$$C_\alpha—C'—\overset{\displaystyle H}{\underset{\displaystyle O}{\overset{|}{\underset{\|}{N}}}}—C_\alpha$$

tend to be planar (that is, all the atoms lie in one plane) because an ethylene-like structure

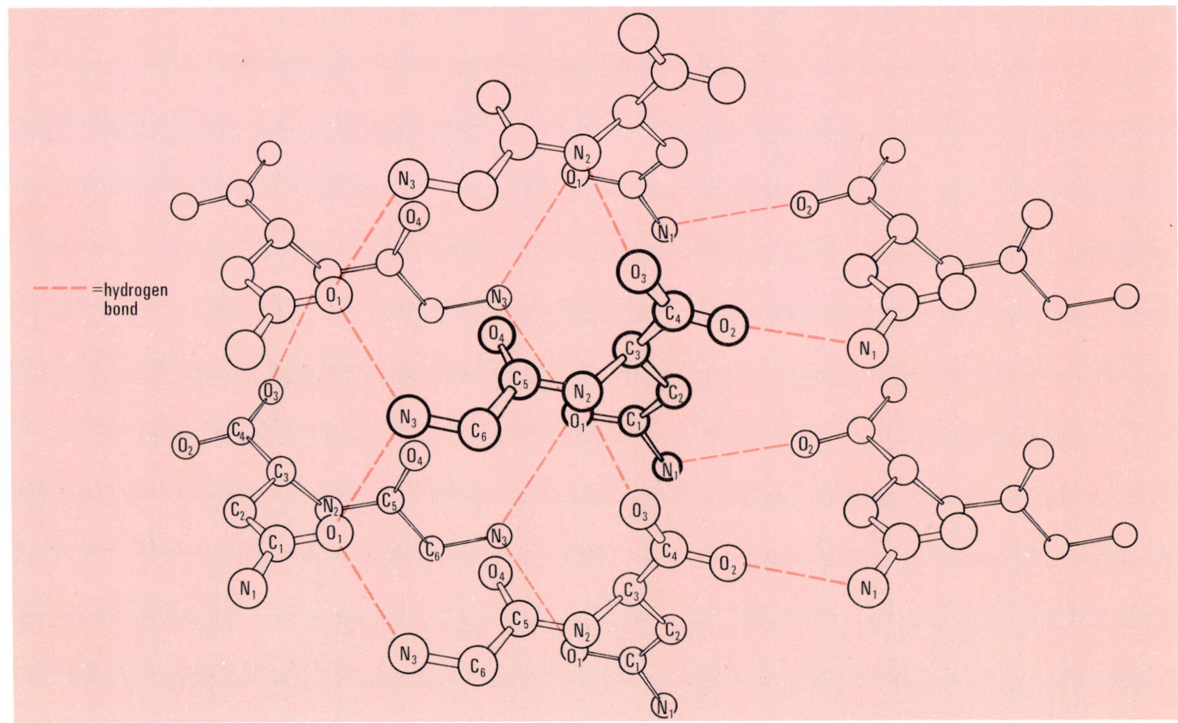

FIG. 3. The crystal structure of the dipeptide glycyl-L-asparagine. One molecule is shown by heavy lines, and the diagram shows how it is surrounded by others in a crystal.

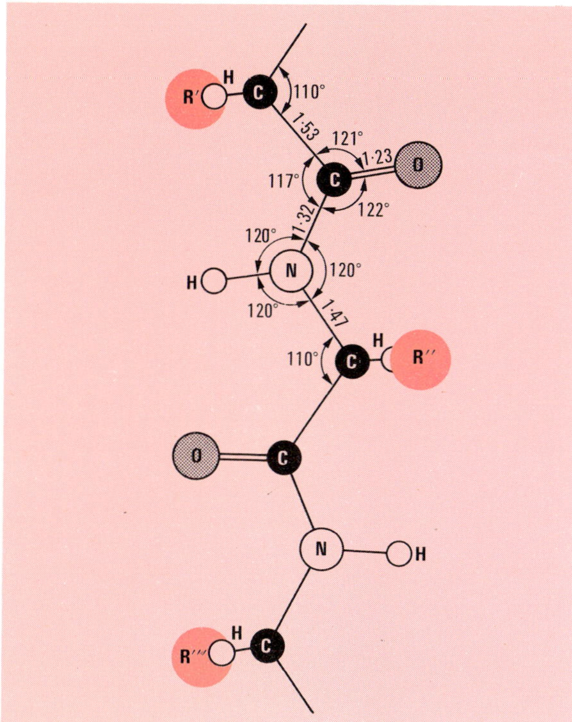

FIG. 4. The dimensions of the peptide group, the geometrical unit from which polypeptide chains are built up, as deduced from studies of individual amino acids and simple peptides.

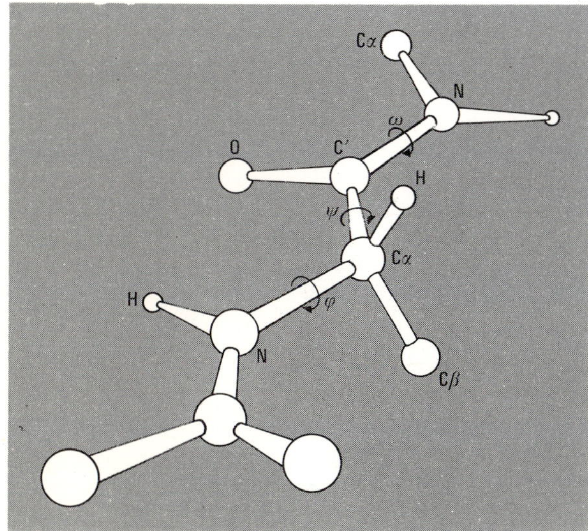

FIG. 5. A dipeptide, comprising two planar peptide groups joined by a tetrahedral alpha-carbon atom. The only geometrical flexibility lies in rotations about the N-Cα and Cα-C' bonds, indicated by the angles ϕ and ψ.

$$\begin{array}{c} H \\ | \\ C_\alpha - C' = N^{\oplus} - C_\alpha \\ | \\ O^{\ominus} \end{array}$$

in which the *peptide* bond is a double bond and the carbonyl oxygen and nitrogen atoms are charged also contributes to their shape; (b) that the bond lengths and angles in these peptide groups have fixed values close to those shown in Fig. 4; and (c) that *hydrogen bonds* can be expected to link the main-chain carbonyl ($>C'=O$)—oxygen atoms to the nitrogen atoms of different peptide groups within the same molecule or in different molecules.

These *hydrogen bonds*, examples of which are indicated in Fig. 3, require special emphasis. They can be regarded as weak ionic interactions between a hydrogen atom, carrying some positive charge because of its covalent connection with an electro-negative atom such as nitrogen or oxygen, and another electronegative atom carrying some negative charge. The peptide group, particularly in its double charged form, is clearly well suited to making hydrogen bonds

$$>N^{\oplus}-H \ldots {}^{\ominus}O-C<$$

These bonds are weaker than covalent bonds but stronger than the common van der Waals' attractions between non-bonded atoms and, furthermore, they are directional since the interaction is strongest when the hydrogen atom lies on the line joining the two electronegative atoms with which it is associated. Consequently, they play an important part in the interactions within and between molecules, holding non-bonded atoms closer together than do van der Waals' forces and constraining groups of atoms to adopt defined orientations.

Armed with their conclusions from studies of amino acids and simple peptides Pauling and Corey considered the possible conformations of poly-peptide chains that would satisfy their criteria of peptide-group planarity and optimum hydrogen bonding. Given a planar peptide group no rotation about the peptide bond is possible, so the flexibility of a polypeptide chain must arise from the possibility of relatively free rotations about the N—C$_\alpha$ and C$_\alpha$—C' bonds at each α-carbon atom (Fig. 5). Detailed analysis shows that these are not completely free rotations, since the two sections of the molecule that are connected by any of these bonds are brought impossibly close together at some settings of the torsion angles, but a wide enough range of angles is accessible to allow a variety of conformations at each α-carbon atom.

Pauling and Corey concerned themselves particularly with structures in which the conformation at each C_α is the same and they examined all such structures in a search for those in which good hydrogen bonds are made. Two general schemes of hydrogen bonding were considered:
(a) with the bonds linking peptide groups that are relatively close together in the same polypeptide chain; and
(b) with the hydrogen bonds linking peptides in different polypeptide chains or widely separated in the same chain.

From these studies two standard conformations of fundamental importance emerged, the α-helix, and the β-pleated sheets.

The α-helix

In the right-handed α-helix the polypeptide chain is coiled up like the thread of a right-handed screw so that hydrogen bonds of optimum length and direction are made between the $\geq$N—H group of amino-acid residue n and the carbonyl oxygen of residue $(n-4)$, that is four residues nearer to the amino end of the chain (Fig. 6(a)). Further studies have shown that this structure is favoured not only by the hydrogen bonding but also by the way in which all of the atoms fit neatly together, like the pieces of a jig-saw puzzle, to make the best possible van der Waals' contacts. The related left-handed α-helix (Fig. 6(b)) is less stable because the C_α and carbonyl-oxygen atoms approach one another too closely. Similarly, helices with hydrogen bonds between different pairs of residues have been considered but calculations show that they are all less stable than the right-handed α-helix. Experimental studies have confirmed this conclusion and have shown that right-handed α-helices are common in proteins of all classes whereas only minor examples of the others have been observed. In particular the right-handed α-helix is the principal conformation of the polypeptide chains in α-keratin, the fibrous proteins of which skin and hair are mainly composed.

Now we can look for the first time at the relationship between the chemical constitution and the three-dimensional structures of proteins since it is clear that some side chains fit well into the α-helix while others do not. The most obvious example is proline which cannot be involved in an α-helix since it has no main chain $\geq$N—H to help form a hydrogen bond. The structure of proline is

$$
\begin{array}{c}
H \\
| \\
H\text{—}N\text{—}C\text{—}COOH \\
\diagdown\ \ \ \ \diagup \\
CH_2\ \ \ \ CH_2 \\
\diagdown\ \ \diagup \\
C \\
H_2
\end{array}
$$

and the dipeptide glycyl proline, for example, lacks the $\geq$N—H (imino) group:

$$
\begin{array}{c}
H\ \ \ \ O\ \ \ \ \ \ \ \ H \\
|\ \ \ \ || \ \ \ \ \ \ \ \ \ | \\
NH_2\text{—}C\text{—}C\text{—}N\text{—}C\text{—}COOH \\
|\ \ \ \ \ \ \ \ \diagdown\ \ \ \diagup \\
H\ \ \ \ \ \ \ CH_2\ \ CH_2 \\
\diagdown\ \ \diagup \\
\underbrace{}\ \ \ \ \ \ C \\
\text{glycyl}\ \ \ \ \ \ \ H_2 \\
\text{residue}
\end{array}
$$

More subtly it appears that the side chains of valyl and isoleucyl residues (Fig. 1), are rather too bulky to fit comfortably in neighbouring positions on the surface of an α-helix. Furthermore some polar side chains, such as those of seryl, asparaginyl and aspartyl residues, can adopt conformations in which they make hydrogen bonds with main-chain groups which weaken or replace the bonds characteristic of the α-helix. On the other hand, the side chains of leucyl, alanyl, glutamyl, methionyl residues, among others, pack together comfortably around the α-helix and tend to stabilize it. Here we begin to see how the amino-acid sequence of a protein molecule might control its overall shape or conformation.

We can note also that fibres made of α-helices might be expected to be elastic to some extent, since we can imagine them recovering from small extensions, but yielding to higher tensions which pull apart the hydrogen bonds that hold the coils of the helix together. The polypeptide chains would then be pulled out into a conformation in which they are much more fully extended. Such a transition is observed in keratin, and Astbury called the two extreme structures the α- and β-forms of keratin. As we have seen, the α-helix (hence its name) is the underlying structure of the α-form. Pauling and Corey's second standard structure accounts for the β-form.

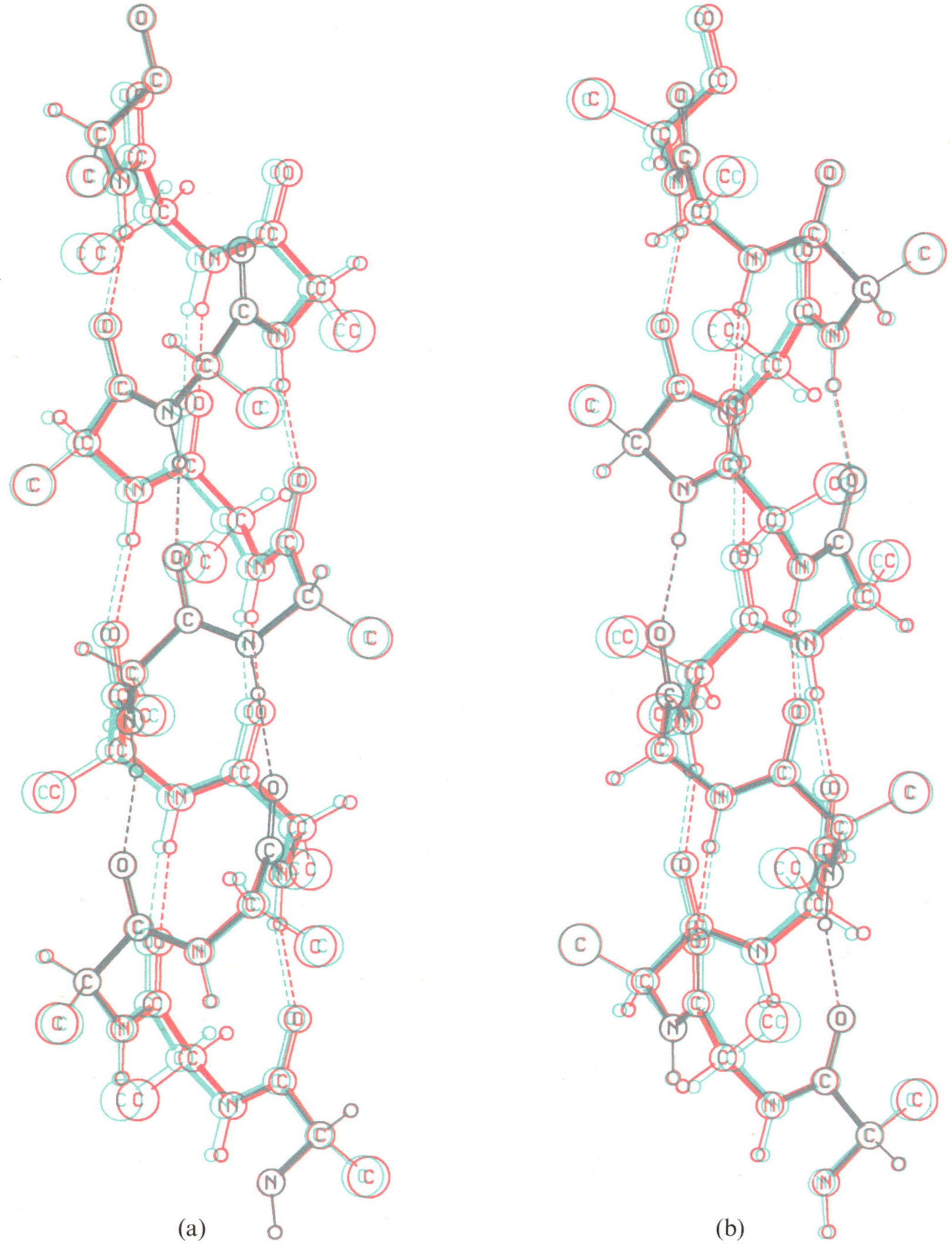

FIG. 6. (a) The right-handed α-helix, (b) the left-handed α-helix. Note that the two are almost, but not quite, mirror images of each other. The drawings show chains in which all the amino acids are alanine, whose side chain is a methyl (CH₃) group. In (a) this side chain is next to the hydrogen atom of the imino group, in (b) it is next to the carbonyl-oxygen atom, which is larger. (See general note on p. 11.)

9

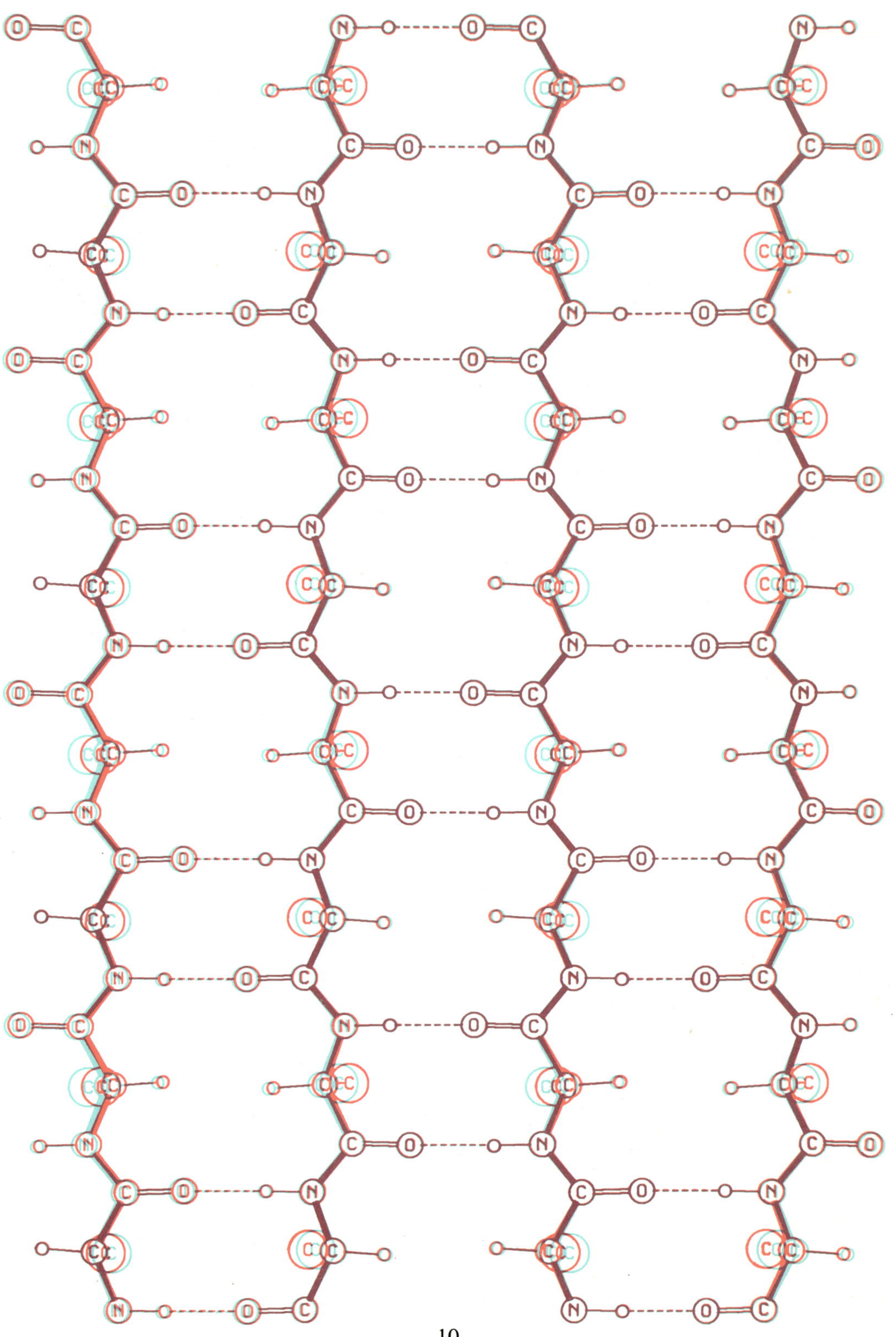

The β-pleated sheets

In the β-structures the polypeptide chain is nearly fully extended with its peptide groups tilted alternately up and down with respect to the mean direction of the chain (Fig. 7). The hydrogen bond donors and acceptors now point outwards, roughly at right angles to the run of the chain, in such a way that good hydrogen bonds can be made to a second chain alongside the first one. These two chains may run in the same direction, in which case the structure is known as a *parallel β-pleated sheet*, or in opposite directions in an *anti-parallel β-pleated sheet*. Both structures are possible and have been observed in fibrous proteins and in globular proteins.

Here again there is a relationship between amino-acid sequence and conformation. When successive sheets in this β-conformation are packed together in layers, as they are in silk, it is necessary for the side chains (which protrude alternately above and below the sheets) to pack together efficiently. This is achieved in the most common silk, from the moth *Bombyx mori*, by the incorporation of a large proportion of glycyl, alanyl and seryl residues in the repeating sequence (Gly-Ser-Gly-Ala-Gly-Ala)$_n$. These amino acids all have small side chains (Fig. 1). In regions with this sequence the glycyl side chains stick out on one side of an individual sheet and the alanyl and seryl side chains stick out on the other side so that adjacent sheets pack together neatly in a highly ordered structure. This structure is disrupted by the inclusion of more bulky side chains, and such breaks in the regular sequence are responsible for occasional disordered regions in the silk fibres which are otherwise mainly well ordered. The fibres are strong and inextensible because any appreciable stretching would require the breaking of *covalent* bonds in the polypeptide chains (cf. the breaking of hydrogen bonds in a stretched α-helix).

GENERAL NOTE. In Figs. 6–9, carbon, nitrogen and oxygen atoms are labelled C, N and O. The large balls labelled C represent methyl (CH$_3$) groups. The small unlabelled balls represent hydrogen atoms. The hydrogen atoms have been omitted from the polyproline drawings, apart from those of the glycine imino groups, which form hydrogen bonds. Hydrogen bonds are indicated by dashed lines.

FIG. 7. Four chains of an anti-parallel β-sheet; the first and third chains run from bottom to top, the second and fourth from top to bottom. Again, alanine side chains are shown.

Structure of collagen

The structure of collagen, the other main class of fibrous proteins, is rather more complicated and was not worked out in quite the same way by a systematic study of the possible conformations. In this case the importance of the amino-acid composition was recognized first and the structure of collagen, which has a very high proportion of glycyl and prolyl residues, was arrived at by analogy with the structures of poly-glycine and poly-L-proline. Many different workers were involved including Cowan, McGavin and North in London; Rich and Crick in Cambridge; and Kartha and Ramachandran in Madras.

Both poly-glycine and poly-L-proline can form helical structures in which there are three amino-acid residues in one full turn of the helix and the molecules aggregate in a very regular three-dimensional manner, forming a crystalline array in which each chain is surrounded by six others. In poly-L-proline (Figs. 8(a) and (b)), where there are no free >N—H groups to form hydrogen bonds,

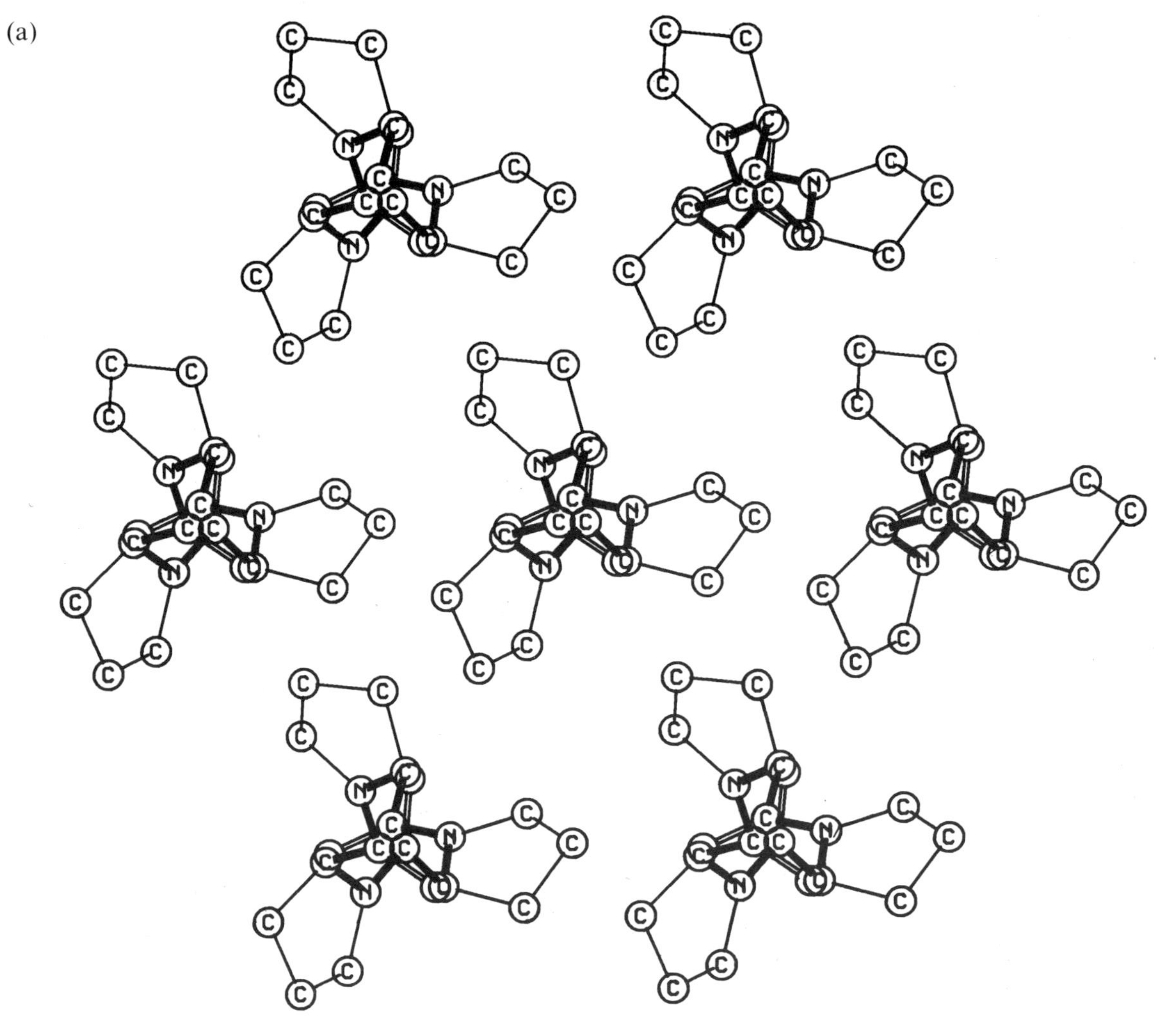

the molecules cohere because they pack together very well; in polyglycine, in addition, hydrogen bonds can form between donor groups and acceptors which point out at right angles to the helix axis. If now we suppose that three neighbouring chains of polyglycine or poly-L-proline are isolated and then twisted together like a rope, we have the basis of the collagen structure though, in detail, this is somewhat more complex. In collagen (Fig. 9) the imino group ($>$N—H) of every third amino-acid residue in each of the three chains is hydrogen bonded to a carbonyl in one of the other chains. The α-carbon atom of this same residue lies close to the axis of the triple helix where there is no room for a side chain, so that this must always be a glycyl residue. The side chains and imino groups of the other two residues in the repeating unit of each

FIG. 8. Two views of the structure of poly-L-proline.
(a) Shows a 'plan' view looking along the length of the chains and shows how they are packed together.
(b) Shows a side view of a set of three chains, the nearest of which is drawn with thicker lines.

(b)

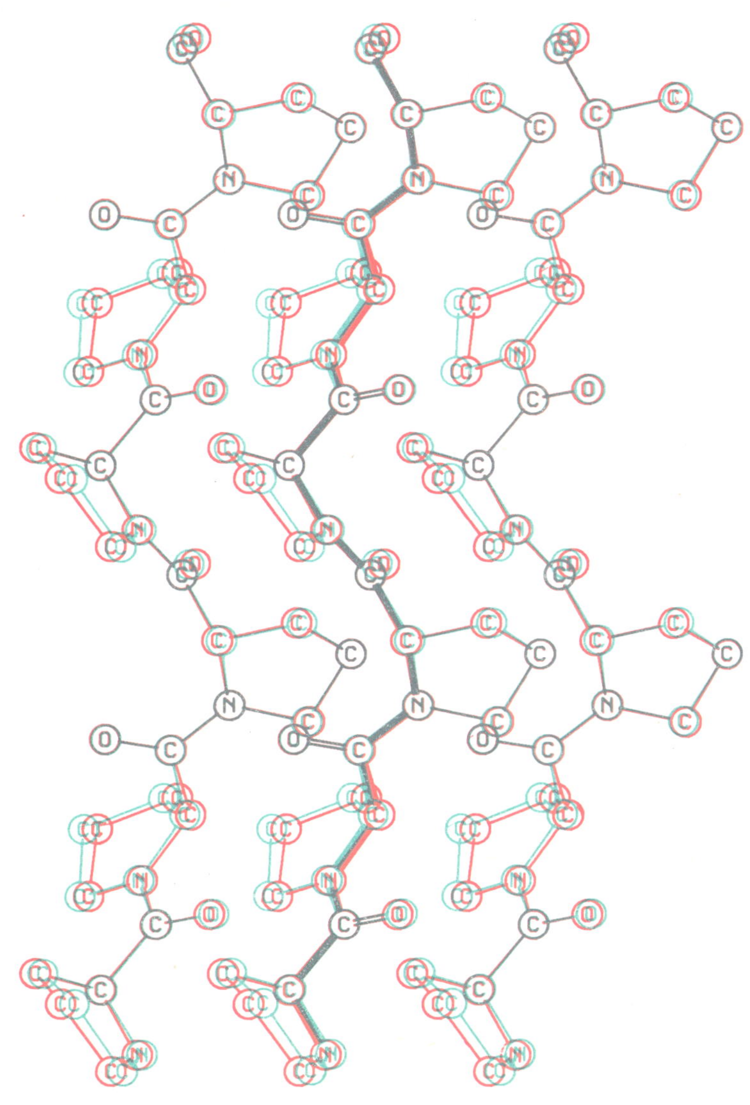

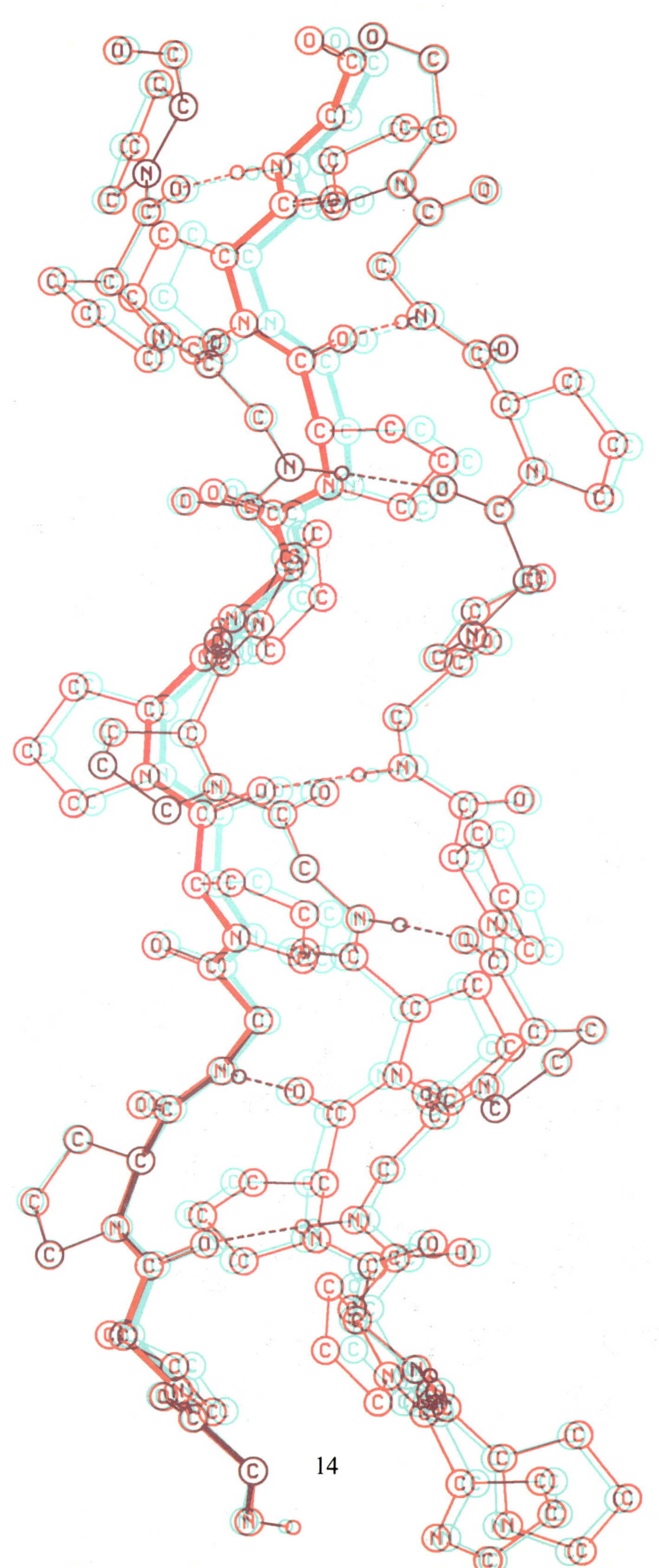

chain point away from the helix axis where there is room for a side chain. Formation of the collagen structure is promoted when at least one of these residues is proline and the molecule has the repeating sequence —(Gly-Pro-X)$_n$— or —(Gly-Pro-Pro)$_n$—, where X is another amino acid.

Like silk, collagen is a strong structure though it can be stretched to a small extent at the expense of deforming or breaking the interchain hydrogen bonds that give it lateral rigidity by linking together the individual polypeptides.

In this brief account of fibrous proteins we have considered only the lowest level of the structure, the arrangement of the atoms in the individual polypeptide chains. The fibres and tissues that we know are built up from very large numbers of these molecules packed together in characteristic ways. We have seen something of how the amino-acid sequence promotes the formation of particular structures at the lowest level and we must note that the sequences also determine how the molecules assemble to form these higher-level structures. Very little is known about this so far but the higher-level structures themselves are being studied intensively and often appear to be various kinds of coiled coils or ropes in which the individual strands are the molecules we have been considering.

GLOBULAR PROTEINS

The first globular protein structure to be worked out in three-dimensional detail was that of myoglobin from the muscle of the sperm-whale. Myoglobin is closely related to haemoglobin and acts as the local store of oxygen in the muscle tissues. In his classic studies, John Kendrew worked on sperm-whale myoglobin because the muscles of these animals are particularly rich in this substance and it can be extracted easily.

In addition to a polypeptide chain of 153 amino-acid residues this myoglobin molecule includes a planar group of atoms with an iron atom at its centre that is called the haem group (Fig. 10). Many globular protein molecules include *prosthetic* groups (of which this is an example) which are usually very closely concerned with the activity of the molecules. In fact the oxygen molecule that is stored by a myoglobin molecule binds to the iron atom in the haem group. The protein part of the molecule provides an environment in which this binding takes place reversibly in a way that is well adapted to the physiological needs of muscle tissue and quite different from the interaction between

FIG. 10. The chemical arrangement of the atoms of a haem group.

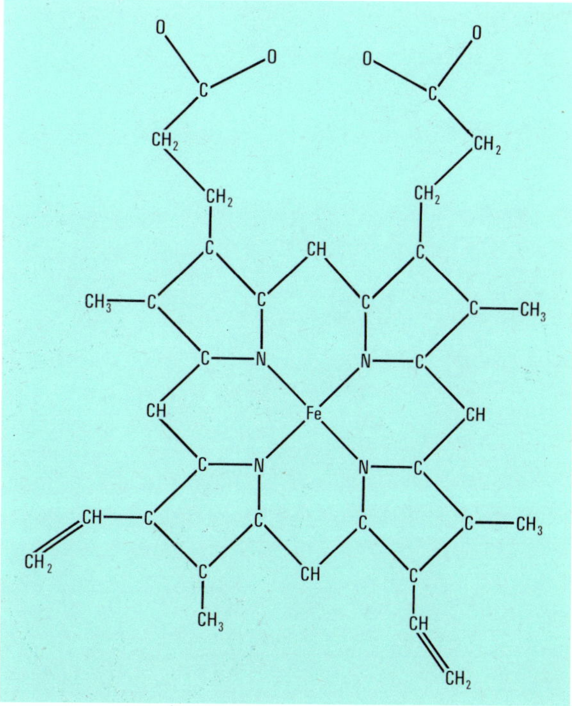

FIG. 9. The structure of collagen may be derived by taking three polyproline chains (as in Fig. 8b), replacing every third residue in each by glycine, and twisting them to form a rope. Although two-thirds of the residues are shown as proline in the drawing, in natural collagen at least half of the proline sites are occupied by other residues.

FIG. 11. Myoglobin, showing the main polypeptide chain and the haem group, top centre. All the side chains have been omitted for the sake of clarity; they would fill in the apparent spaces between different sections of main chain and form the points of attachment to the haem group. Note that the central iron atom of the haem group (the large ball) is not quite in the plane of the other atoms.

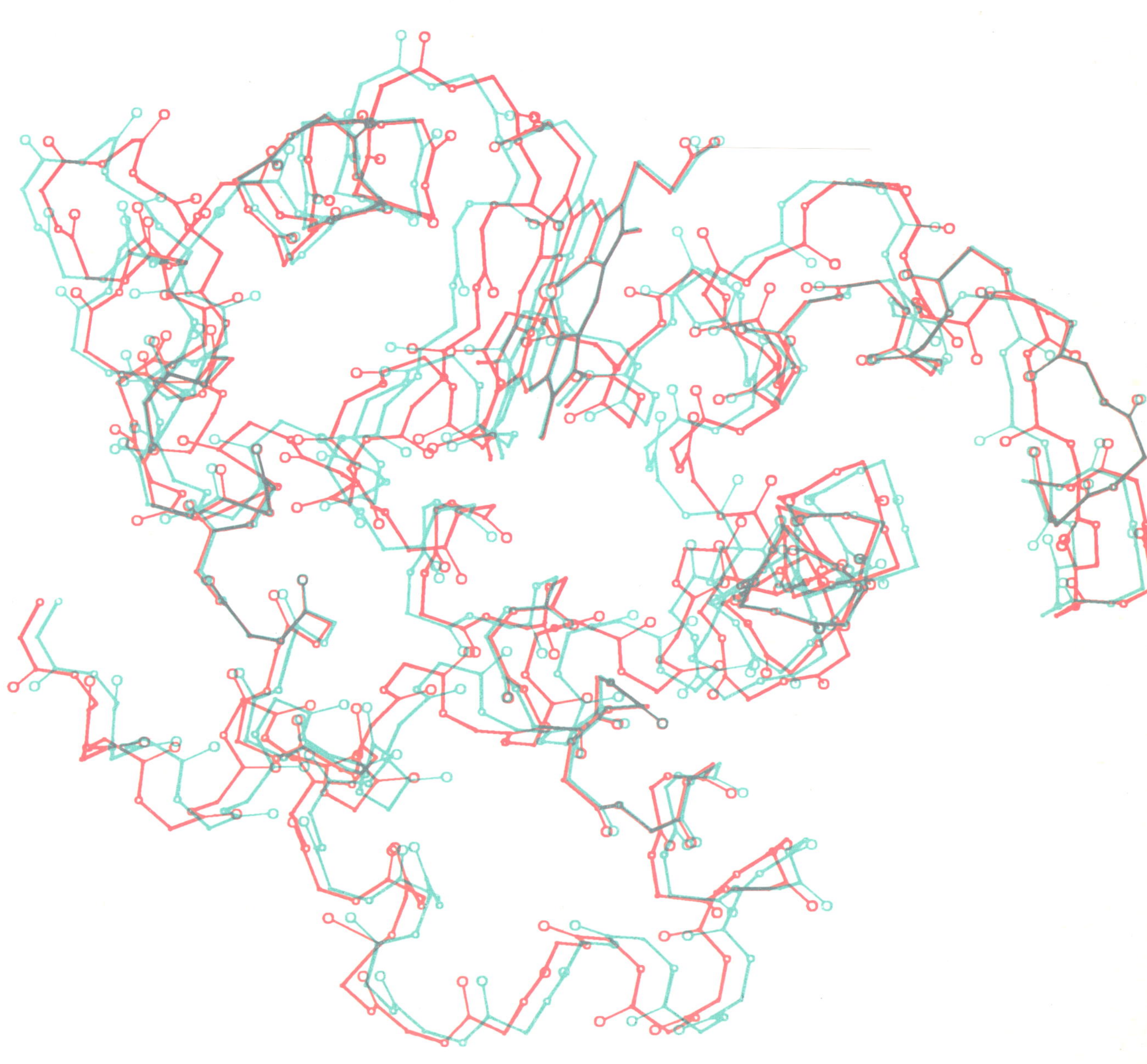

oxygen and free haem in aqueous solution.

Kendrew's crystallographic studies showed that most of the polypeptide chain in myoglobin is in the α-helical conformation with dimensions very close to those predicted by Pauling and Corey. In order to create a nest for the haem group, eight lengths of helix connected by lengths of chain in less regular conformations are packed together in a way that seems, at first sight at least, to be remarkably irregular (Fig. 11). The whole molecule illustrates well the terminology suggested by the Danish protein chemist Linderstrom-Lang in which the amino-acid sequence is called the *primary structure*; the local conformation of the polypeptide chain (here predominantly α-helical) is called the *secondary structure*; and the general folding of the coiled chain is called the *tertiary structure*. For larger molecules composed of more than one polypeptide chain there is a higher level of organization again, the *quaternary structure* which refers to the way in which folded chains or sub-units pack together to form larger aggregates. This level is well illustrated by the structure of haemoglobin. Again by crystallographic analysis, Max Perutz has shown that this molecule is made of four sub-units, which are identical in pairs and closely similar to individual myoglobin molecules, arranged roughly at the apices of a tetrahedron.

Distribution of polar and non-polar side chains

The structure of myoglobin immediately confirmed a connection between the amino-acid sequences of proteins and their three-dimensional structures that had long been suspected. We must remember that some amino-acids (such as threonine where R is $-CH(CH_3)OH$ and asparagine where R is $-CH_2.CO.NH_2$) have side chains which include oxygen and nitrogen atoms and are polar, since they carry partial positive and negative charges through polarization of their electron distributions. Such polar side chains attract water molecules, which are also polar, and form hydrogen bonds with them. Some polar side chains are also ionizable (for example, aspartic acid $-CH_2.COO^\ominus$ and lysine $-CH_2.CH_2.CH_2.CH_2.NH_3^\oplus$) and attract water molecules even more strongly. All of the side chains that attract water are said to be *hydrophilic* while the others with non-polar side chains, (for example leucine,

$$R = -CH_2.CH\begin{cases}CH_3\\CH_3\end{cases})$$

do not ionize, do not attract water molecules, and are said to be *hydrophobic*. Inspection of the molecular structure shows that the different kinds of amino-acid side chains are distributed unevenly throughout the structure: the molecular surface is composed mainly (though not exclusively) of the hydrophilic polar and ionizable side chains while the interior of the molecule is largely non-polar and hydrophobic in character. As Langmuir and Rideal had proposed such molecules are well described as 'oil drops with polar coats'.

In order to appreciate the reason for this we must remember that such a molecule takes up its active conformation and functions in an aqueous environment: in considering how it folds up we must take into account the interactions between the protein and the surrounding water molecules. Following Kauzmann the current theory is that non-polar side chains, such as those of leucyl or phenylalanyl residues, are surrounded by an ordered network of water molecules when they are exposed to the solvent. Burying them in a regularly folded structure shields them from the water which is then much less regularly arranged—at the cost of selecting one conformation of the polypeptide chain from the many that are accessible to the unfolded molecule. (A thorough thermodynamic analysis of the folding process shows that the decrease in entropy due to the selection of one folded conformation for the protein molecule is compensated to some extent by an increase in the entropy of the surrounding water.) The balance of opposing influences seems to be finely poised for most proteins so that they can be unfolded or denatured by relatively small changes in their environments, such as increases in the temperature (as in boiling an egg) or the addition of agents like urea to their solutions. An important part of the evidence that the chemical constitution of a protein fixes its three-dimensional structure comes from observations that these molecules will often fold again spontaneously to their active conformations when normal conditions are restored after denaturation.

Consideration of the interactions between proteins and water calls into question the importance of hydrogen bonds in determining protein structures since water itself provides an excellent source of hydrogen bond donors and acceptors: surely the imino and carbonyl groups of the main chain and the hydrogen bonding groups of the polar side chains can bond to water molecules when the

Fig. 12. Lysozyme, showing the main polypeptide chain, but omitting all the side chains. (Compare with Fig. 13 (e) in which the side chains are included.)

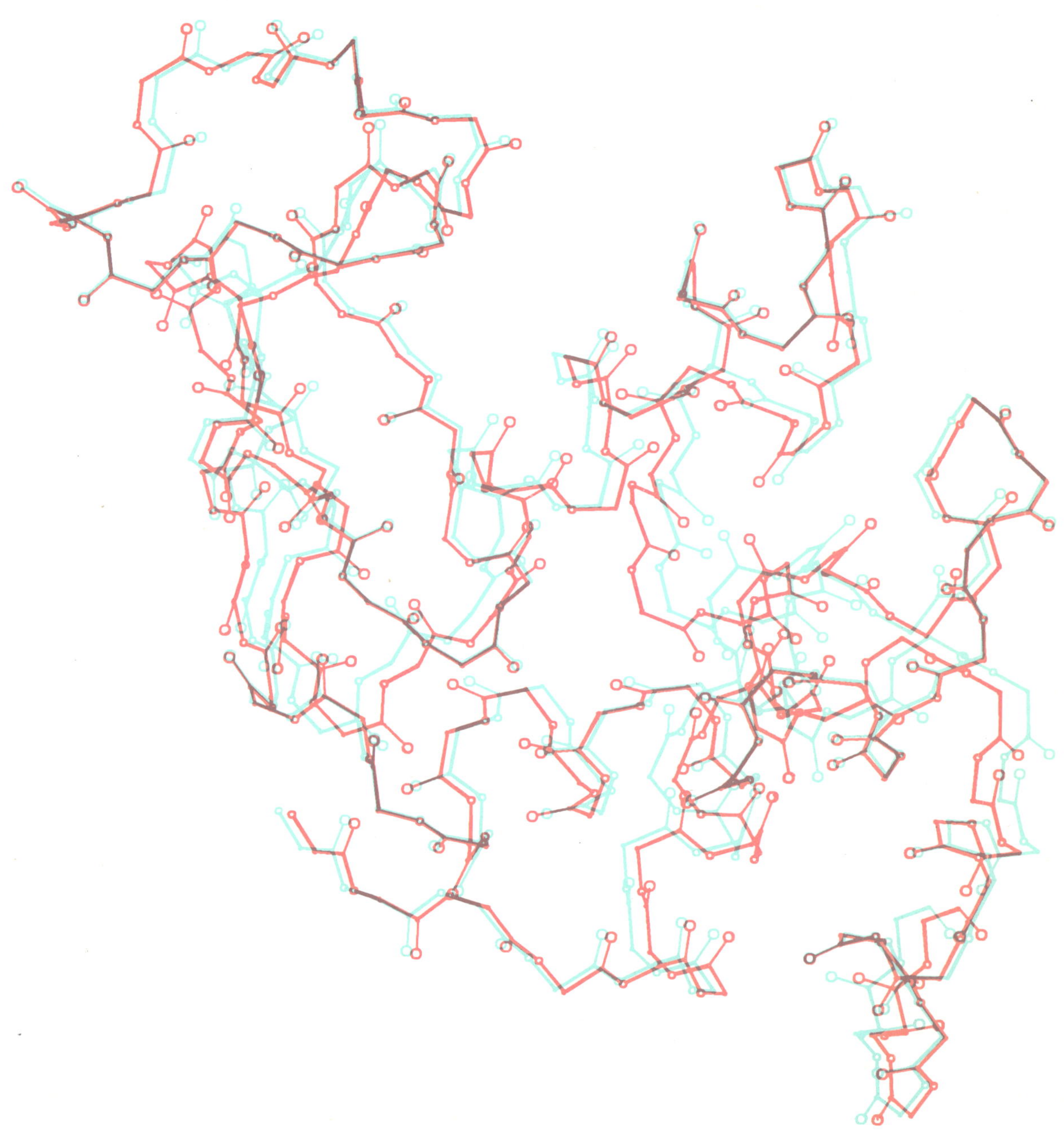

18

protein is unfolded and there can be little if any energetic advantage in their making similar bonds within the protein molecule itself. This objection appears to be a sound one and the main impulse in folding these molecules must come from the more general effects of the interactions with solvent that have just been mentioned. Nevertheless hydrogen bonds remain important in fixing the folded structures of protein molecules because various hydrogen bonding groups are bound to be shielded from the liquid in any folded structure and it is then most important energetically that they should be able to make at least as good bonds within the molecule as they made previously to the water. The directional properties of these bonds then help to fix the relative positions and orientations of groups within the folded molecule with the precision that is needed for their function.

Lysozyme

Despite its seeming irregularity the structure of myoglobin provided a good starting point for the examination of globular proteins since more recently determined structures have proved to be even more complex. By now about a dozen more structures are known in similar detail but there is no room to discuss them all here. Most of them are enzymes and the first of these to be worked out was the structure of *lysozyme*. This substance, which is found in tears, mucus and various other body fluids and in especially large quantities in the white of eggs, was discovered by Alexander Fleming some years before his discovery of penicillin. It is also an anti-bacterial agent, but, unlike penicillin, it is an enzyme and acts by promoting the hydrolysis of covalent bonds that hold together the molecules which make up the walls of certain bacterial cells. Its three-dimensional structure and probable mode of action were worked out by the authors and their colleagues at the Royal Institution in London between 1962 and 1966.

The hen egg-white lysozyme molecule (Fig. 12) is roughly egg-shaped with dimensions about $3 \times 3 \times 4$ nm. About 40% of the amino-acid residues are in α-helices, some of which are rather irregular, while a further 12% of the residues make up a region of anti-parallel β-structure. Although the general arrangement of the polypeptide chain seems considerably more complex than that found in myoglobin there is a pattern to it that is not difficult to remember and which may indicate how the chain

folds up.

From experiments involving the radioactive labelling of haemoglobin molecules growing in red blood cells and of lysozyme molecules synthesized in the chicken oviduct, it is known that when a protein is being synthesized at the ribosome, the amino end of the chain is made first: residues are added one by one towards the carboxyl end while the part of the chain already synthesized pushes out from the position on the ribosome at which the synthetic reactions take place. Little is known about the environment of the growing chain but it is an attractive idea that proteins begin to fold up to their active conformations as they are being made. The structure of lysozyme fits this hypothesis very well, and it is most easily described as though it folded in this way, though the actual course of events for this or any other protein must not be supposed to be as straightforward as this simple description might suggest. In particular, each part of a growing polypeptide chain does not take up immediately the conformation that is found in the finished molecule. Extensive rearrangements must occur as more of the chain is synthesized and the most that we dare hope may be that, for some proteins at least, well-defined parts of the growing chain fold first, act as centres for the rest to fold around and survive, recognizably, in the completed molecule.

In the hen egg-white lysozyme molecule, starting at the amino end of the polypeptide chain, we find (Fig. 13) that residues (1–4) are in a more or less extended conformation leading to the first length of α-helix (5–15). Residues (16–24), in a complex conformation with some traces of β-folding, join this first helix to a second less regular one which includes residues (25–36). The two helices lie across one another with their axes roughly at right angles and with a number of non-polar side chains in contact between them, forming what looks like a plausible centre of folding for the rest of the molecule. Residues (37–41) are folded back towards the amino end of the chain—making hydrogen-bond contact with it—and they are followed by a run of polar residues (42–54) which form a hairpin-like section of anti-parallel β-pleated sheet which is not in contact with the preceding part of the folded chain but forms a wing pointing away from it. It is tempting to suppose that these hydrophilic side chains remain happily in contact with the surrounding liquid and fold back on themselves to form the hairpin in order to bury the hydrophobic residues (55–56) in the non-polar pocket formed by the earlier residues.

The succeeding residues (57–86) are of mixed types and fold on the hairpin β-structure in such a way that up to residue 86 the polypeptide chain forms a structure with two wings shaped rather like the letter V. Part of the gap between the two arms is filled by another length of somewhat irregular helix (87–98) in which non-polar side chains occur at the appropriate intervals to fall inside the molecule in contact with others in the existing hydrophobic pocket. Finally residues (99–129) are folded around the nucleus formed by the first forty residues.

In this account no mention has been made of the four disulphide bridges (—S—S—) which are made, one supposes, as the individual cysteine residues are brought together appropriately by the general folding of the chain. Certainly it is difficult to imagine any long-range interactions between these residues that might pull the chain towards its observed conformation but there is no doubt that the disulphide bridges endow the finished molecule with additional stability.

It remains to be seen whether model schemes of this kind lead to an effective method of predicting the folded structures of protein molecules from their amino-acid sequences. Much work is in progress on these lines and there has been some success in identifying sequences that fold into α-helices: much remains to be done, especially perhaps in establishing the role of water in the folding process and showing how it can be taken into account in calculations.

FIG. 13. The lysozyme molecule growing by stages.
(a) The first 15 residues; the first 6 residues, starting at
the left hand, form an extended chain, like one strand of a
β-sheet; then follow three turns of an α-helix running
nearly vertically upwards.

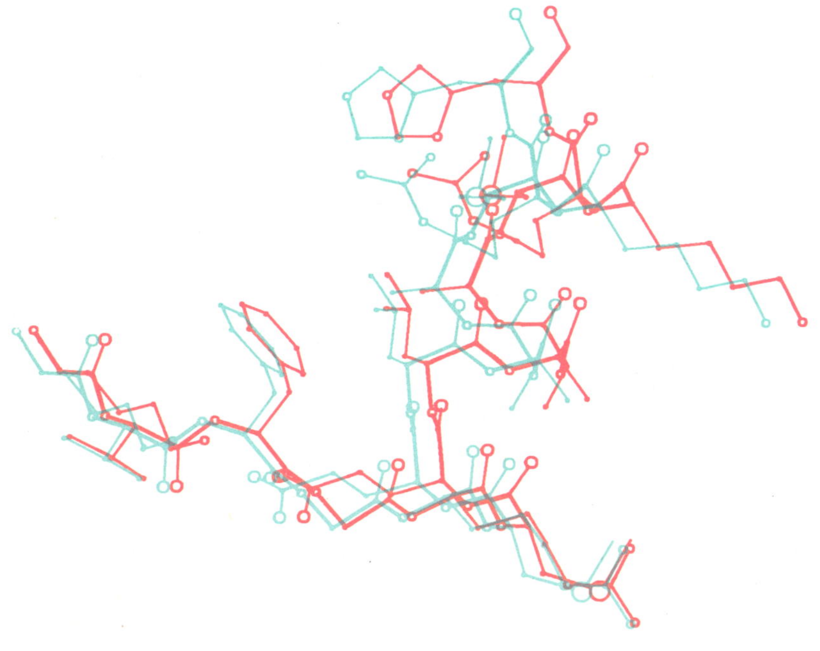

21

(b) The first 38 residues; residue 15 is followed by a rather
irregular stretch, then by another three-turn helix running
roughly horizontally from right to left.

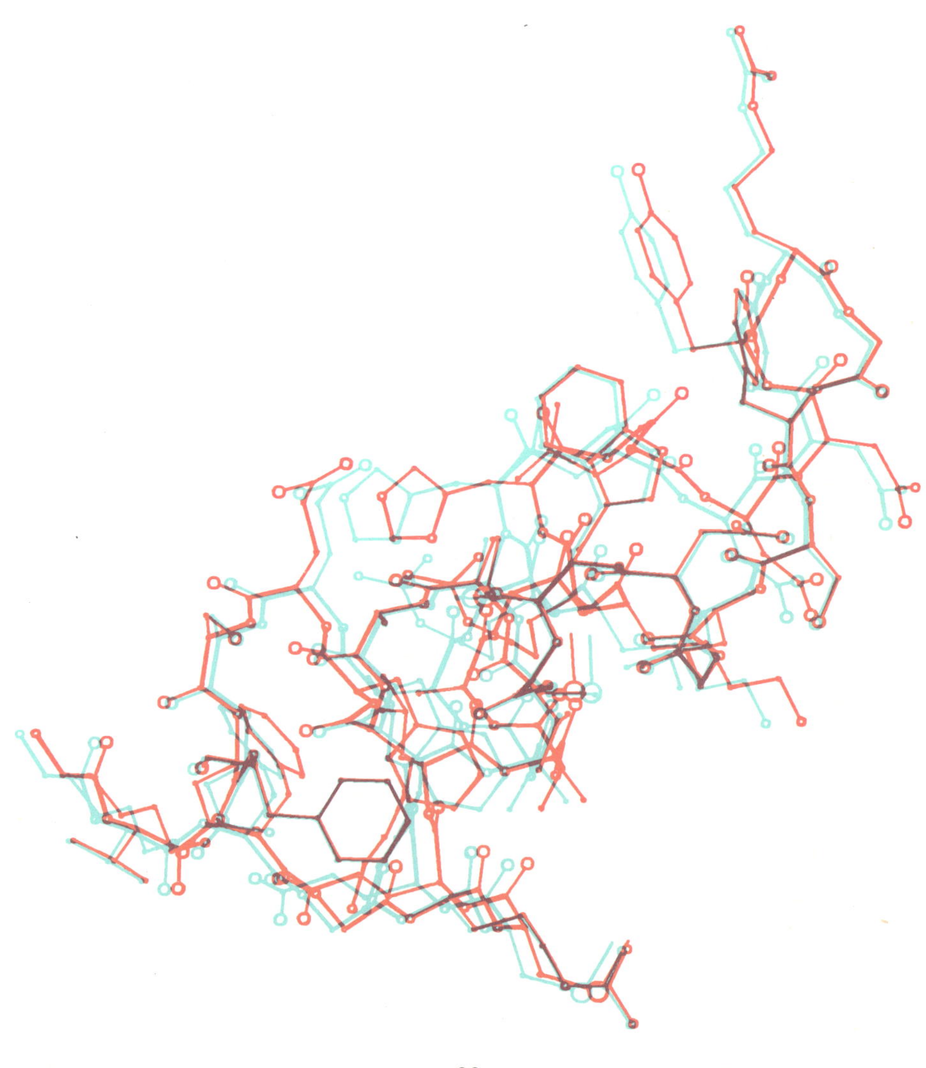

(c) The first 61 residues; following residue 38, the chain runs up to residue 49, doubles back down to 55 and then up again to 61. These three stretches of extended chain form three strands of a rather irregular anti-parallel β-sheet. The molecule now looks like the two limbs of a letter V.

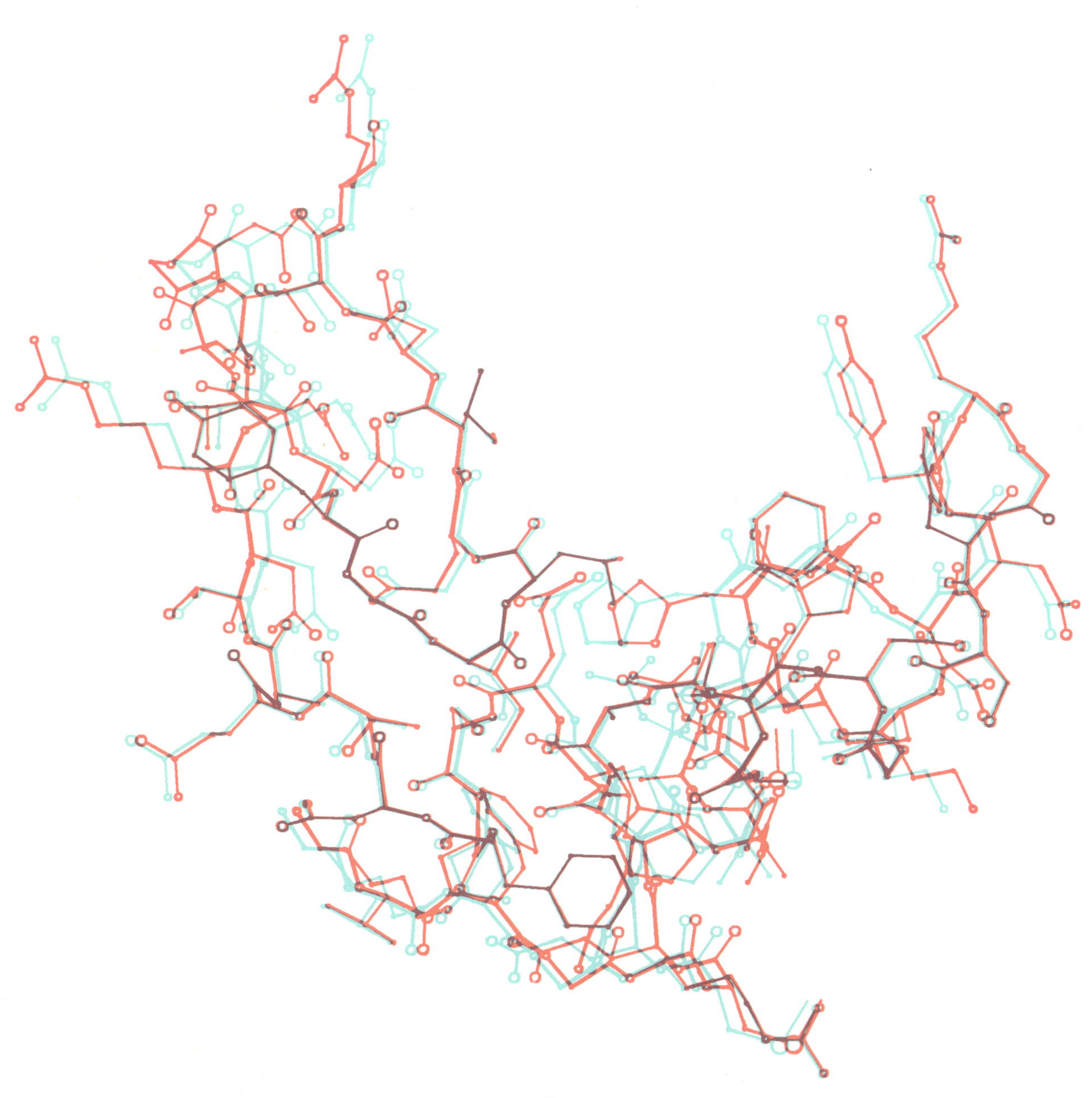

(d) The first 100 residues; the residues added at this stage form first a rather irregular arrangement behind residues 39–61 and then a third length of α-helix running upwards to the right and forwards, largely filling in the gap between the limbs of the V.

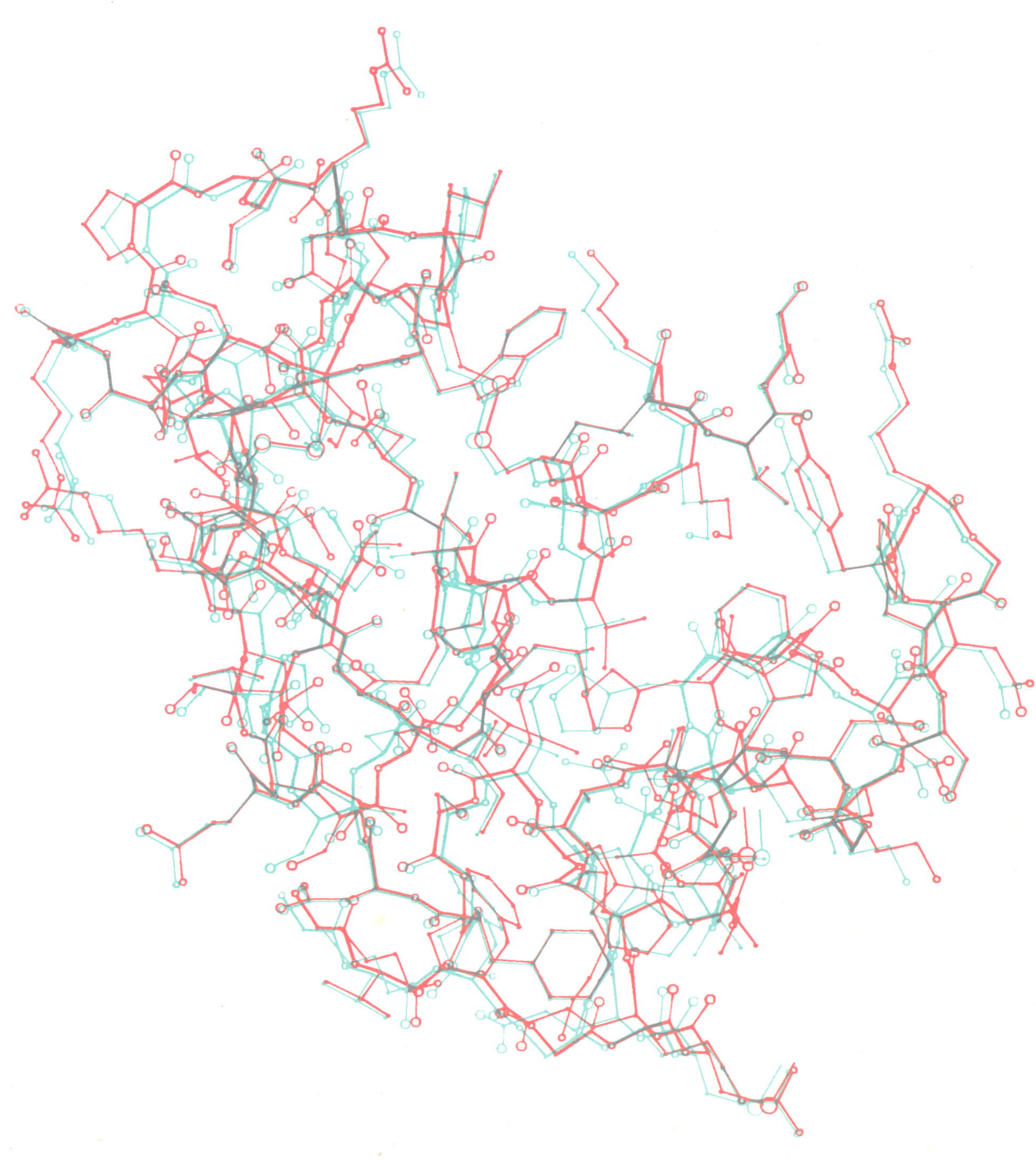

(e) Finally, the last 29 residues form an additional strand over the right-hand and bottom-right part of the molecule. Of course, the lysozyme molecule does not really have a 'top' and a 'bottom'; molecules can be any way up in solution, but it is convenient for descriptive purposes to select some suitable arbitrary orientation.

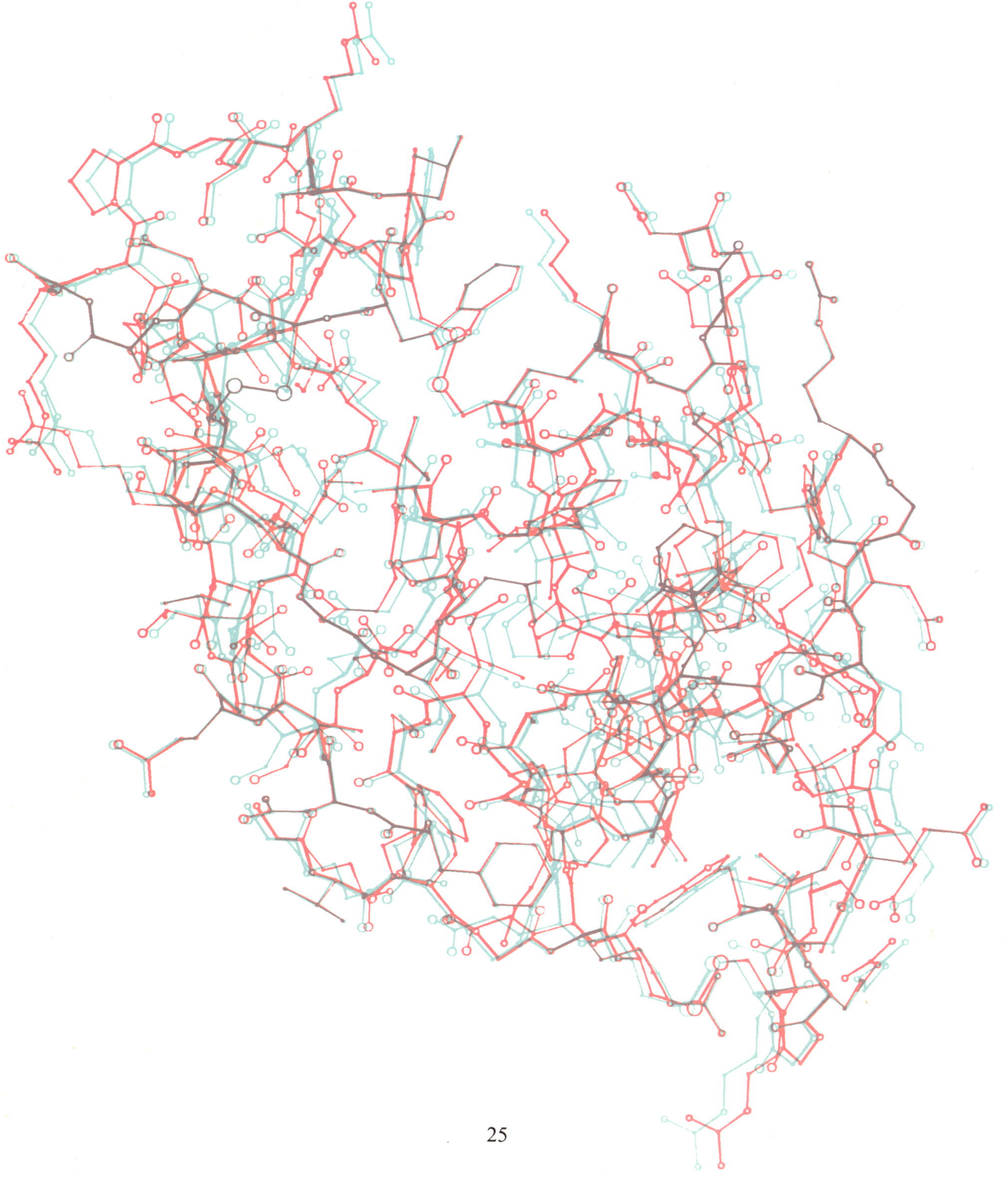

Interaction of enzyme and substrate

The most marked feature of the lysozyme molecule is a deep groove that cuts across it (Figs. 12, 13e). This is the active site of the enzyme, shown in more detail in Fig. 14, where the interaction with substrate takes place that leads to the cleavage of a glycosidic bond (the connecting bridge, through an oxygen atom, between two sugars) at a greatly increased rate. The mechanism of catalysis does not concern us here but we must note that all enzymes appear to be proteins, some having prosthetic groups of various kinds, and that they interact with their substrates through the same kinds of weak bonds that are important in stabilizing the enzyme structures themselves. The rate of virtually every chemical reaction, however simple, that takes place in living cells is controlled by an enzyme, and lysozyme, the example we have examined in detail, is one of the very smallest of them. Many of the most important are very large molecules composed, like haemoglobin, of various numbers of sub-units and the forces that hold these sub-units together again involve the same weak interactions and depend intimately upon the structure of the surrounding water. The principle of self-assembly appears still to apply : the structures of these molecules also are determined by their chemical structures. Similarly we suppose that this same principle governs the assembly of the great protein organelles, such as muscle, which are built up of protein molecules of various kinds—some fibrous, some globular—having complex enzymic activities.

FIG. 14. The heavy lines indicate six sugar molecules linked together, forming part of a polysaccharide chain, one of the constituent molecules of a bacterial cell wall. The lighter lines represent those parts of the lysozyme molecule that form the active site cleft, into which substrate molecules are bound. The hydrogen bonds formed between enzyme and substrate are shown. The various amino-acid types are labelled according to a standard 3-letter code.

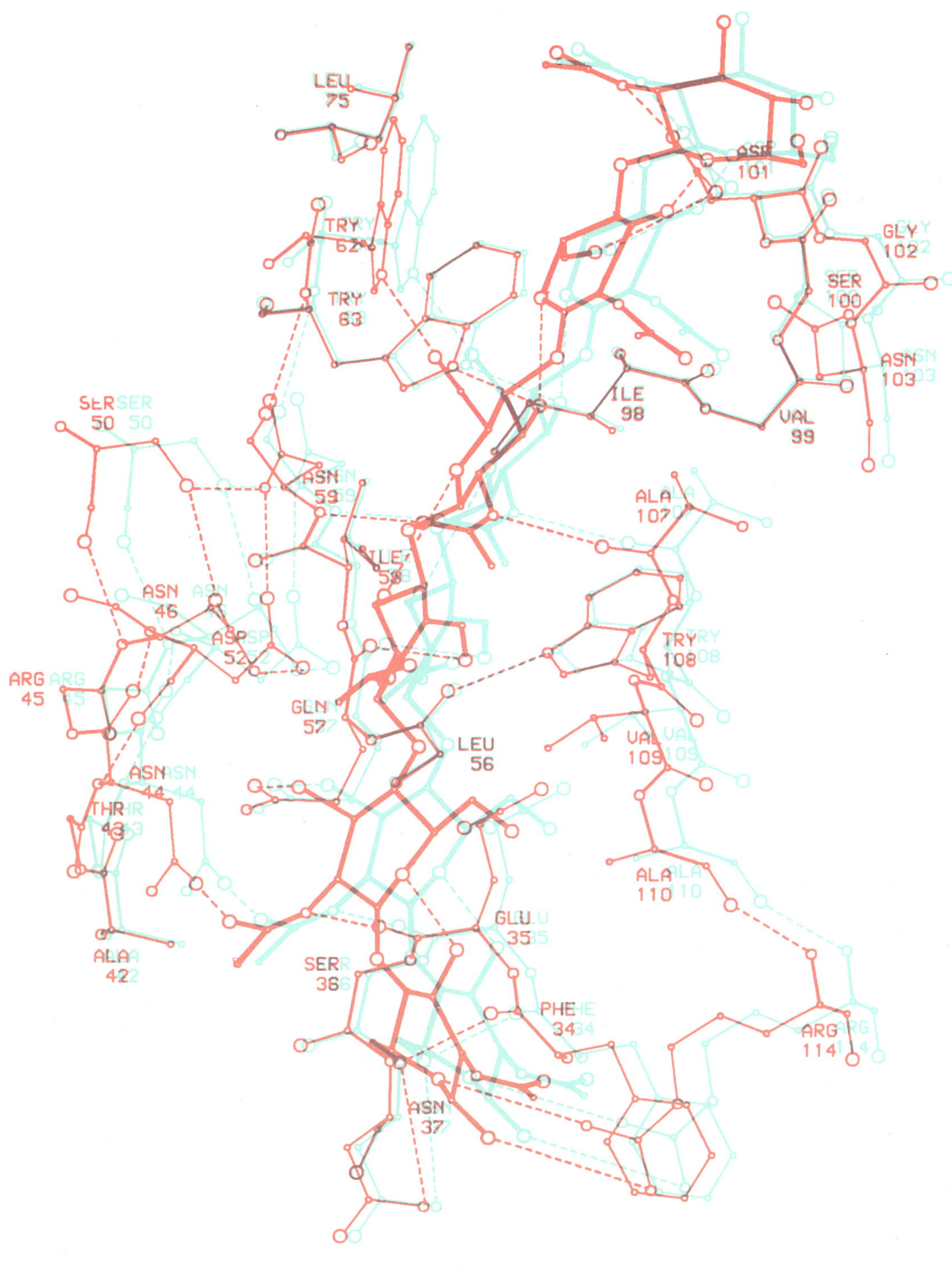

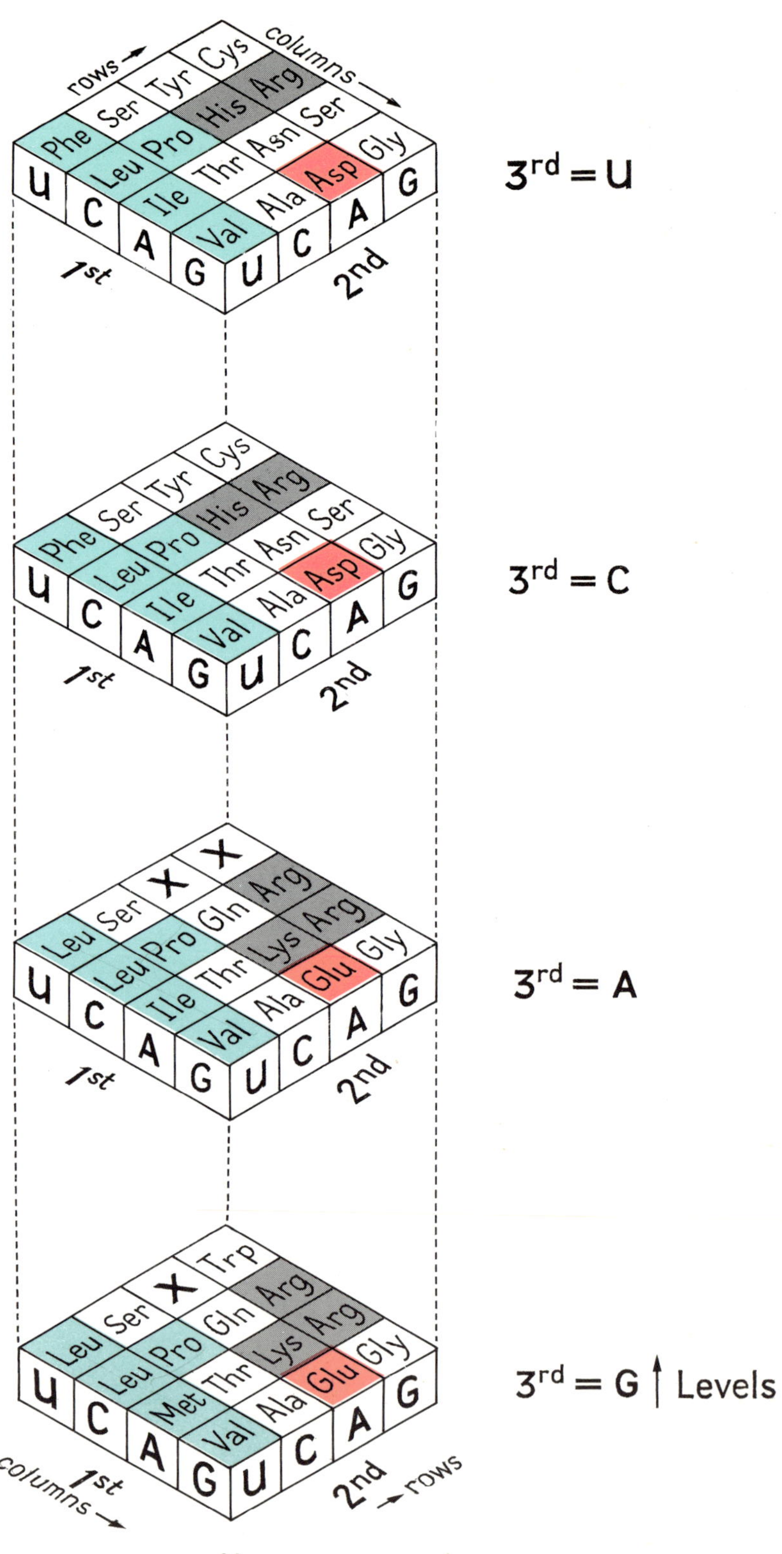
3rd = U
3rd = C
3rd = A
3rd = G Levels
1st
2nd
rows
columns

Blueprints for protein assembly

Let us then consider for a moment what determines the chemical structure of a protein and how the multitude of different protein molecules has come into existence. Here we find at once that the proteins are wrongly named since a different type of macromolecule more truly holds first place. These are the nucleic acids of which the genes are made that pass the design of every organism from one generation to the next: the nucleic acids are like the blueprints for a factory which is built and occupied by proteins. In a way that has been discovered in detail during the last ten years, the nucleic acids carry in code the chemical structures of the proteins.

The nucleic acids that control protein synthesis are also long-chain polymer molecules composed of alternating phosphate groups and sugar molecules each of which carries as a side chain one of four different nitrogenous bases: adenine, guanine, cytosine and uracil. The genetic information is encoded in the order of these bases which determines the order of the amino-acid residues in the corresponding protein molecules. The relationship between the two is known as the *genetic code*. It is a triplet code in which three consecutive bases in the nucleic acid—a codon—(which can be chosen in $4 \times 4 \times 4 = 64$ ways) code for one amino acid residue. Since there are more triplet codons available than there are amino-acid types, several different codons usually code for each amino acid and there are some left over for the punctuation marks to show the beginning and end of each polypeptide chain (Fig. 15).

This is not the place to describe in detail all that is now known about protein synthesis and the genetic code but two features of the latter are especially relevant to our subject.

First, changes in the nucleic acids that affect the nature of particular side chains will give rise to corresponding changes in individual amino-acid residues in the proteins for which they code. This is an important mechanism of evolution: specific changes in the nucleic acids lead to specific changes of particular amino-acid residues in the corresponding protein molecules. In the course of time this may lead to profound changes in the protein molecules, but they take place residue by residue. Such changes can be observed by comparing proteins with the same functions that have been isolated from different species: thus in human insulin, residue B 30 at the free-carboxyl end of the

B chain is threonine instead of the alanine that is found in pig insulin (Fig. 2). Otherwise the two molecules are chemically identical.

Second, inspection of the code shows that codons corresponding to amino-acid residues of the same general type (for example valine and leucine where

$$R = -CH_2 \Big\langle \begin{matrix} CH_3 \\ CH_3 \end{matrix} \quad \text{and} \quad -CH_2-CH \Big\langle \begin{matrix} CH_3 \\ CH_3 \end{matrix}$$

respectively, see Fig. 1) tend to be similar and differ generally by only one of the three bases. The importance of this constraint is seen when we consider the structures of proteins less closely related than human and pig insulins. Human and hen egg-white lysozymes differ widely in their amino-acid sequences yet they have been shown to have very closely similar three-dimensional structures. Since the shape of the molecules depends so intimately upon the distribution of the polar and non-polar side chains this must mean that changes in sequences have a better chance of proving acceptable if at least the character, polar or non-polar, of the residues is conserved. Such conservatism apparently is built into the code.

Finally, how have radically different proteins evolved, given the existence of some primitive organism to start with? Here a comparison of amino-acid sequences suggests a possible answer. Thus it has been discovered that human lysozyme and bovine α-lactalbumin, a milk protein with a different function, have related chemical structures. We might imagine, therefore, that in some primitive vertebrate the lysozyme gene was accidentally duplicated, one copy was conserved for the production of lysozyme but the other, relatively free to change without detriment to the organism, eventually found a new function in the evolution of the mammals in coding for α-lactalbumin.

Findings of this kind are beginning to provide information about evolutionary relationships at the molecular level and they suggest that the many varied protein molecules belong to a limited number of families, the members of which are recognizably similar in structure though not always in function. If this is so the problem of predicting the three-dimensional structure of a protein molecule from its chemical sequence may be transformed, as more structures are worked out, into the problem of identifying the evolutionary family to which the molecule belongs. Thus we may soon uncover a pattern of proteins that will reveal in detail how so much has been achieved with units of such simple chemical structure.